The Power

of

Unified Physics

By

Isaac Lasley

The Power of Unified Physics

Isaac Lasley

Lulu.com Edition

This book was written to help improve the public understanding of unified physics to open the door to many new technological resources. The primary sources of information come from research scientist Isaac Lasley's creative imagination and abstract reasoning, but all are based upon proven scientific facts.

The Power of Unified Physics

Introduction

All matter is composed of electromagnetic particles, waves, fields, and energy flow. This book describes the applications of unified physics technologies proven to be possible in the book **Gravity Now!!**, to show ways in which some of the specific technological breakthroughs of truly unified physics will function to improve the future of humanity, as it has now been proven that the basis for everything, including the universe, is pure positive and negative electromagnetic energy! The electromagnetic particles being the proton and the electron each have their own waves, fields, and energy. Electromagnetic energy flows within and between electromagnetic particles, waves, and fields.

Humans created the sundial!! Time Is A Man Made Conception! Time is a concept created as a tool to better understand life and the lifespan, on earth. Humans created the names for the seasons, a calendar to keep track of days during each year, and the hours on the clock to keep track of the sun or moon during days which would all continue to exist without names for them. Earth (24 hour day) time does not exist in outer space, as biological time has been proven to exist on planet earth as a result of planetary electromagnetic energy, beneath the earths crust and above in the earth's atmosphere. Biological age is the result of the electromagnetic energy on the cells to do cellular work on a magnetic level due to the energy of gravity in the atmosphere pushing down from above and pulling down on us from beneath. Gravity exists on earth as a result of the electromagnetic energy that exists on both a planetary and atomic level. The electromagnetic energy that is the basis of all things in the universe is evident by understanding chemistry down to the proton and electron, the neutron is simply a proton and electron, as well as the fact that the proton and electron exist on the electromagnetic spectrum as do all things in the universe. All matter exists as electromagnetic particles, waves, fields, and energy flow!

The earth revolves around the sun. Its electromagnetically based wobble around our sun is due to the earth and the sun's energy fields and energy flow which causes the earth's daily, and seasonal changes as a result, since time does not physically exist in outer space. The planetary systems rotate through our Milky Way galaxy as a result of the fact that the electromagnetic energy continuously flows through our galaxy energy fields flow around our galactic electromagnetic black hole, similarly to electrons around the nucleus an atom. The galaxies rotate through our universe as the electromagnetic energy fields and continuous energy flow through the universe. The universe may or may not be expanding, as our perspective points of view of other planetary systems and galaxies are likely continually changing as a result of the electromagnetic fields and continuous energy flows on both galaxy and universe macroscopic levels. This technology should not be very hard to understand considering that even ants can clearly communicate using radio frequencies to expand their communities to serve there queen for the greater good of the ant populations.

Unified physics understands that all levels of physical science are based on the fact that all levels of existence result from electromagnetic energy particles, waves, fields, and energy flow. Chemistry is based on elements that are composed of pure electromagnetic energy particles known as protons and electrons each with their own electromagnetic waves, fields, and energy flow. Biology is based on the study of cells that are made of chromosomes which are created from DNA/RNA that are composed of electromagnetically bound particles made of chemical elements.

6

Unified physics have great potential for the progress of humanity, the health of our earth's environment, and the later intergalactic progress of humanity. Tesla coils of many different sizes will likely be needed to help power every innovative device noted in this book.

Remember!

A Relatively Flat "Space-Time Continuum" Has Been Proven False And "Space-Time" Is Only Used For Distance Measurement In Outer Space!

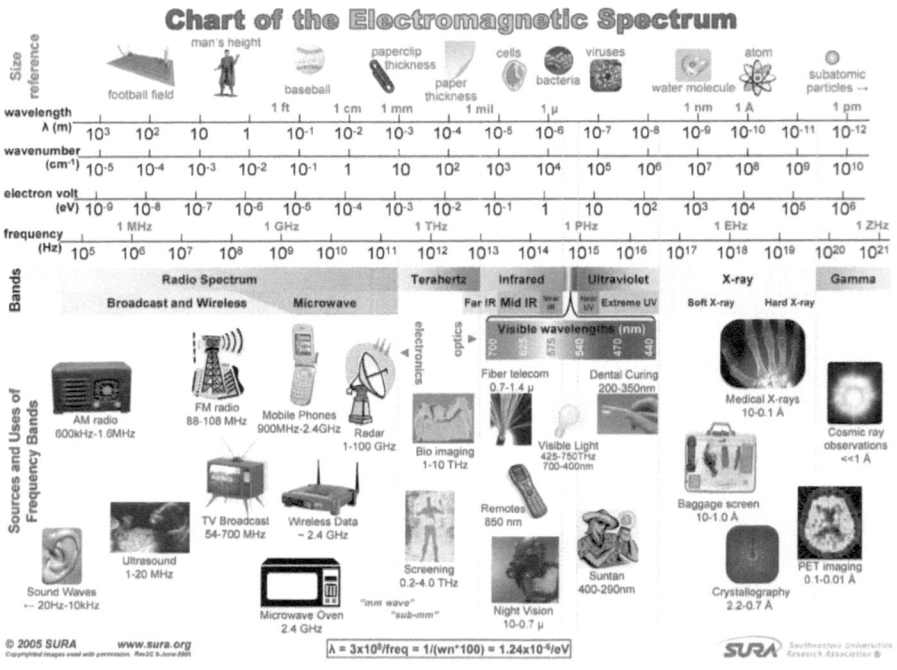

Unified Physics

Biological time exists on earth as a result of the work that cells do as a result of gravity. Gravity occurs as a result of the electromagnetic layers of atmosphere pressing down above us and the electromagnetic based earth's core pulling down beneath us. Everything contained within, and including the universe, has now been proven beyond reasonable doubt to be electromagnetically bound. The atom, biological systems, planetary systems, solar systems, galaxies, and the universe itself all function on an electromagnetic basis. All matter and antimatter are composed of electromagnetically charged particles, waves, fields, and energy flow that exist on the electromagnetic spectrum with all types of electromagnetic energy. Energy equals mass times the speed of light squared ($E=mc^2$). Since energy equals mass times the speed of light squared then the mass equals energy divided by the speed of light squared ($m=E/c^2$) and the speed of light squared equals energy divided by mass ($c^2=E/m$). All of these equations make electromagnetic replication and disintegration absolutely possible, as well as teletransportation.

The nucleus of an atom contains both protons and electrons, but not neutrons, only positive and negative electromagnetic particles of energy with electromagnetic waves, fields, and energy flow for each of them. The "neutron" is the electromagnetic particle equivalent of an alternating current, while the proton and electron are the positive and negative electromagnetic particle equivalent of a direct current, and a "neutron" is clearly composed of merely a proton and electron. As this undeniable fact has now been realized we should strive to understand this new reality of physical sciences to effectively create a better future for all of humanity. The atom, planetary systems, solar systems, galactic systems, and the universe are all electromagnetically based. All electromagnetic energy is connected in a pattern of formation for particles, waves, fields, and energy flow on both microscopic and macroscopic levels. The replication processes of transcription and reverse transcription are electromagnetic processes used by DNA and RNA to reproduce and modify cells on a genetic level.

DNA is one of the electromagnetic code bases of biological life on earth. Deoxyribonucleic acid [DNA] has backbones made of

10

deoxyribose sugar [C5H10O4] and phosphate [PO4] with gene sequences made of nucleotides adenine, guanine, thymine, and cytosine. Adenine and guanine are both purines that form pairs bound to thymine or cytosine which are both pyrimadines. DNA is formed into a double helix chain with a positive (p) and negative (q) strand. As a result of the electromagnetic energy flow through the chain of purines bound to pyrimadines by electromagnetic hydrogen bonds. Nucleotides are all composed of hydrogen, carbon, nitrogen, and oxygen. Adenine is the purine that binds to the pyrimadine thymine in DNA molecules. The chemical formula for adenine is C5H5N5. Guanine is the purine that binds to the pyrimadine cytosine in DNA molecules. The chemical formula for guanine is C5H5N5O. Cytosine is the pyrimadine that binds to the purine guanine in DNA molecules. The chemical formula for cytosine is C4H5N3O. Thymine is the pyrimadine that binds to the purine adenine in DNA molecules. The chemical formula for thymine is C5H6N2O2.

RNA is another electromagnetic code basis for life. Ribonucleic acid has backbones made of ribose sugar [C5H10O5] and phosphate [PO4] with gene sequences made of nucleotides adenine, guanine, thymine, and cytosine. Adenine and guanine are both purines that form pairs bound to thymine or cytosine which are both pyrimadines. RNA is formed into a double helix chain as a result of the electromagnetic energy flow through the positive and negative side of the chain of purines bound to pyrimadines by hydrogen bonds. Nucleotides are all composed of hydrogen, carbon, nitrogen, and oxygen. Adenine is the purine that binds to the pyrimadine uracil in RNA molecules. The chemical formula for adenine is $C_5H_5N_5$. Guanine is the purine that binds to the pyrimadine cytosine in RNA molecules. The chemical formula for guanine is $C_5H_5N_5O$. Cytosine is the pyrimadine that binds to the purine guanine in RNA molecules. The chemical formula for cytosine is $C_4H_5N_3O$. Uracil is the pyrimadine that binds to the purine adenine in RNA molecules. The chemical formula for uracil is $C_4H_4N_2O_2$.

Genes are heredity units composed of noted groups of DNA that determine certain cell structures and functions. Since genes are basically composed of various patterns of DNA they obviously contain carbon, hydrogen, nitrogen, and oxygen. Chromosomes are composed of groups of genes, as every human has twenty three pairs of chromosomes. Since chromosomes consist of various patterns of genes; they are also simply composed of carbon, hydrogen, nitrogen, and oxygen.

Cells are the biological foundation of life, as each cell is life within a larger self. Every cell in all plants and animals has a life and death of its own. Biological cells tend to reproduce more than once during their lifetime within plants and animals. Every cell within each organism contains smaller intricate organelles with unique functions that work together with the goal of homeostasis on a cellular level, down to single celled organisms. With this understanding of life's intricate processes we will be able to replicate or teletransport them simply by better understanding their electromagnetic code or signature.

Life Elements

Hydrogen is the element of the universe that truly proves that the basis of unified physics is absolutely true and sound beyond any doubt. Hydrogen naturally occurs on earth as a diatomic gas molecule that is electromagnetically neutral, as a single atom has an atomic weight around 1. It is likely so abundant throughout the universe since it is so electromagnetically basic and pure. A pure hydrogen atom is composed of one positive electromagnetically charged proton particle, electromagnetically bound with one negative electromagnetically charged electron particle to create one electromagnetically neutral atom with electromagnetic particles, waves, fields, and energy flow! Of course hydrogen will likely be the first atom created using a replicator since it is so basic and likely be one of the easiest tests to initially perform.

Although, we must remember that the hydrogen "ions" of deuterium and tritium will exist between hydrogen and helium on a new electromagnetic periodic table. Deuterium has an atomic weight around 2 while still retaining a neutral charge, meaning that it contains 2 protons and 1 electron in the nucleus, with 1 electron in the electromagnetic field orbiting the nucleus. Tritium has an atomic weight around 3 while still retaining a neutral charge, which means that the atom contains 3 protons and 2 electrons in the nucleus, with 1 electron in the electromagnetic field orbiting the nucleus. Both deuterium and tritium react differently to other elements than hydrogen as a result of the varying amounts of electromagnetic energy that is present within each atomic nucleus. Hydrogen makes

14

up about 9.5% of the human body's mass. It is a common constituent of most organic molecules, as well as water [H2O]. Hydrogen in its ionized form H+ makes bodily fluids more acidic.

Carbon [C] is one of the primary bases of life on earth as the majority of life forms on earth are carbon based. Carbon occurs naturally on earth as a non metal solid that is electromagnetically neutral, having an atomic weight around 12. A pure carbon atom is composed of 12 protons and 12 electrons. A carbon atom has 6 electrons electromagnetically bound to 12 protons in the nucleus and 6 electrons in the electromagnetic energy fields orbiting the positively charged atomic nucleus with electromagnetic waves flowing between the nucleus and electron fields, as with all atoms. Carbon makes up about 18.5% of the human body's mass. It also forms the backbone like chains of all organic molecules, including lipids [fats], carbohydrates, proteins, and nucleic acids [DNA and RNA].

Nitrogen [N] naturally occurs on earth as a nonmetal gas that is electromagnetically neutral, having an atomic weight around 14. A pure nitrogen atom is composed of 14 protons and 14 electrons. A nitrogen atom has 7 electrons electromagnetically bound to 14 protons in the nucleus, with 7 electrons in the electromagnetic energy fields orbiting the positively charged nucleus. Nitrogen makes up about 3.2% of the human body's mass, as it is a component of all nucleic acids and proteins.

Oxygen [O] naturally occurs on earth as a nonmetal gas that is electromagnetically neutral, having an atomic weight around 16. A pure oxygen atom is composed of 16 protons and 16 electrons. An oxygen atom has 8 electrons electromagnetically bound to 16 protons in the nucleus, with 8 electrons in the electromagnetic energy fields orbiting the positively charged nucleus. Oxygen makes up about 65% of the human body's mass. It also makes up the majority of mass of the water molecule [H_2O]. Oxygen is a part of many organic molecules containing carbon, and is used to generate ATP. Oxygen molecules are also used by cells to temporarily store electromagnetic [chemical] energy within them.

Sodium [Na] naturally occurs on earth as an alkali metal solid that is electromagnetically neutral, having an atomic weight around 22. A pure sodium atom is composed of 22 protons and 22 electrons. A sodium atom has 11 electrons electromagnetically bound to 22 protons in the nucleus, with 11 negatively charged electrons in the electromagnetic energy fields orbiting the positively charged nucleus.

16

Sodium makes up only about 0.2% of the human body's mass, but it is needed to generate action potentials. Ionized sodium [Na+] is the most abundant cation in extracellular fluid, as it is also essential for maintaining proper water balance.

Magnesium [Mg] naturally occurs on earth as an alkali earth metal solid that is electromagnetically neutral, having an atomic weight around 24. A pure sodium atom is composed of 24 protons and 24 electrons. A sodium atom has 12 electrons electromagnetically bound to 24 protons in the nucleus, with 12 electrons in the electromagnetic energy fields orbiting the positively charged nucleus. Magnesium makes up only about 0.1% of the human body's total mass. Ionized magnesium [Mg+2] is needed for the actions of many enzymes, as these molecules increase the rate of chemical reactions [electromagnetic interactions] within organisms.

Phosphorus [P] naturally occurs on earth as a nonmetal solid that is electromagnetically neutral, having an atomic weight around 30. A pure phosphorus atom is composed of 30 protons and 30 electrons. A phosphorus atom has 15 electrons electromagnetically bound to 30 protons in the nucleus, with 15 electrons in the electromagnetic energy fields orbiting the positively charged nucleus.

Phosphorus makes up about 1% of the human body's mass. It is a component of ATP and nucleic acids, as well as being required for normal tooth and bone structure.

Chlorine [Cl] naturally occurs on earth as a nonmetal gas that is electromagnetically neutral, having an atomic weight around 35. A pure sodium atom is composed of 35 protons and 35 electrons. A sodium atom has 18 electrons electromagnetically bound to 35 protons in the nucleus, with 17 electrons in the electromagnetic energy fields orbiting the positively charged nucleus. Chlorine only makes up only about 2% of the human body's mass. Ionized chlorine [Cl-] is the most abundant anion in extracellular fluid as it is essential for maintaining the proper water balance.

Sulfur [S] naturally occurs on earth as a nonmetal solid that is electromagnetically neutral, having an atomic weight around 32. A pure sulfur atom is composed of 32 protons and 32 electrons. A sulfur atom has 16 electrons electromagnetically bound to 32 protons in the nucleus, with 16 electrons in the electromagnetic energy fields orbiting the positively charged nucleus. Sulfur only makes up about 0.25% of the human body's mass, but it is a component of many proteins and some vitamins.

Potassium [K] naturally occurs on earth as an alkali earth metal solid that is electromagnetically neutral, having an atomic weight around 39. A pure potassium atom is composed of 38 protons and 38 electrons. A potassium atom has 19 electrons electromagnetically bound to 38 protons in the nucleus, with 19 electrons in the electromagnetic energy fields orbiting the positively charged nucleus. Potassium only makes up about 0.35% of the human body's mass. Ionized potassium [K+] is the most abundant cation in intracellular fluid and is needed to generate action potentials.

Calcium [CA] naturally occurs on earth as an alkali earth metal solid that is electromagnetically neutral, having an atomic weight around 40. A pure calcium atom is composed of 40 protons and 40 electrons. A calcium atom has 20 electrons electromagnetically bound to 40 protons in the nucleus, with 20 electrons in the electromagnetic energy fields orbiting the positively charged nucleus. Calcium makes up about 1.5% of the human body's mass, as it contributes to the hardness of teeth and bones. Ionized calcium [Ca+2] is needed for many processes in the human body like blood clotting, contraction of muscles, the release of some hormones, and many other functions.

Iron [Fe] naturally occurs on earth as an alkali metal solid that is electromagnetically neutral, having an atomic weight around 55. A pure sodium atom is composed of 55 protons and 55 electrons. A sodium atom has 29 electrons electromagnetically bound to 55 protons in the nucleus, with 26 electrons orbiting in the electromagnetic energy fields surrounding the positively charged nucleus. Iron only makes up about 0.005% of the human body's mass, but it does play an important part in distributing electromagnetic [chemical] energy throughout the human body. Ionized iron atoms [Fe+2 and Fe+3] make up an important part of hemoglobin protein to carry oxygen in red blood cells, as well as being a part of some enzyme proteins that catalyze electromagnetic [chemical] reactions is living cells.

Trace Elements make up about 0.2% of the human body's mass as they may include: Aluminum [Al], Boron [B], Chromium [Cr], Cobalt [Co], Copper [Cu], Fluorine [F], Iodine [I], Manganese [Mn], Molybdenum [Mo], Selenium [Se], Silicon [Si], Tin [Sn], Vanadium [V], and Zinc [Zn].

Every biological organism, like all matter, is composed of a combination of chemicals which are ultimately composed of pure electromagnetic energy particles, waves, fields, and energy flow. With this understanding it can simply be concluded that all matter, as it is composed of pure electromagnetic energy, can also be electromagnetically manipulated and modified. The science fiction technologies of the replicator, disintegrator, teletransporter, and vehicles that can travel many times faster than light now no longer have to be works of fiction as they can now be physically created. Humanity and our planet earth can be vastly improved upon in the very near future, which is undeniably necessary after a twentieth century where humanity destroyed and polluted the planet earth and everything upon it at a faster rate than any other century before. New York existed for more than one hundred years without smog and the people who currently live in such highly populated areas can enjoy life without smog in the near future. We can begin to enjoy a planet where future generations will no longer need to live on a planet where fossil fuels are the primary source of energy and the internal combustion engine is our primary means of transportation. We currently utilize electromagnetic energy on many levels of medical diagnostics and treatments in magnetic resonance imagining [MRI],

CT scan, x-rays, electron microscopes, and lasers.

<u>Replicator</u>

Basics

A replicator will be relatively easy to create as unified physics are now proven to show that all things are composed of electromagnetic energy particles, waves, fields, and energy flow. Mass equals energy divided by the speed of light squared ($m=E/c^2$). Technically a replicator will be considered a digitally programmed electromagnetic integrator that is made to replicate anything that is scanned or programmed into its database of electromagnetic product codes. Every element on the periodic table has been proven beyond any doubt to be composed of pure positive and negative electromagnetic energy particles, waves, fields, and energy flow. Since all matter is composed of elements, all matter is therefore composed of pure electromagnetic energy.

Humanity has been stimulating atoms to create electricity from negatively charged electrons while ignoring the protons from which they are also composed for more than a century. Now it is time to utilize both the proton and electron electromagnetic energy to begin creating matter of any kind at our will. As a result of replication technology the food production businesses, logging industry, product production factories, retail stores, and many other product based businesses will likely have to adapt to industry changes. This technology will help us to be sure that the production costs of many products can drop, while profits soar. This technology will also change the medical treatment industry forever to make wounds heal more quickly and efficiently that ever before regardless of the quality of a patient's immune system. Viruses can replicate after infecting a cell, humans should be able to replicate nearly anything we want, and now we are closer to creating a replicator than ever.

I have noted how many basic chemical elements can be formed using this technology by noting their electromagnetic compositions. The specific chemicals that I have noted are all vital to life on some level and are relevant to tissues that are noted by their electromagnetic composition later in this chapter. By noting how a replicator could be utilized to create matter, specifically biological cell tissues in this chapter, we can clearly see how it can become very useful to humanity on many levels when it is created, as it is clearly possible to design. Right now we can create pure positive and negative electromagnetic energy called electricity. We have split the hydrogen atom to observe the great nuclear power exerted from separating the pure electromagnetic bond within it. We have isolated both the proton and electron, as well as created a "neutron", as positive and negative electromagnetic energy is attracted to the opposing charge.

WE NOW LIVE IN THE 21ST CENTURY!!!! We Need To Start Acting Like It By Creating Technologies Now Proven Possible! We need to begin using our control over electromagnetic energy particles, waves, fields, and energy flow to create atoms and matter by creating replicator technology that is now proven to be possible through unified physics. Starting a new age of scientific revolution we should probably follow with the creation of a new "periodic table" that includes ions. Ions exist as a result of varying amounts of the charged particles, waves, fields, and energy flow from which they are composed, which is what makes them react to other elements so much differently than their base element. Ions are in essence different elements which are clearly apparent as they are each composed of varying amounts of electromagnetically charged particles. Cations and anions commonly form electromagnetic ionic bonds, as positive and negative electromagnetic energy attract. Other elements with varying numbers of electrons in their outer fields tend to find other elements to form covalent bonds to share electrons to make the outer fields contain eight electrons together. Everything, including the universe, is composed of electromagnetically bound particles, waves, fields, and energy fields of pure positive and negative electromagnetic energy with waves, fields, and energy flow existing between particles.

The General Design of Replicator Units

Most basically a replicator will have an electromagnetic scanner, a hard drive, and an electromagnetic particle, wave, field, and energy flow integrator that functions to replicate items previously scanned after having their electromagnetic signature. The most challenging part of creating a replicator will be gaining power over the proton to integrate with the electrons available from electricity. Making the specific electronic hardware and software that will be necessary to make the scans to note the size, location, and consistency of all electromagnetic particles, waves, fields, and energy flow which is being replicated will be the easy part. A scan would then be followed by a second scan to insure that the digital electromagnetic map of item or subject that is being replicated is created properly for every particle, wave, field, and energy field before electromagnetic replication process is started. After the digital electromagnetic map is created and stored on the replicator hard drive with a code or file it can be selected at a later point to be replicated. Since replicators will in essence digitize anything that can then be replicated at a later point and time, grocery and retail stores could become digitized to some extent on the internet for immediate replication download. The first replicators will likely be primarily used to replicate only foods.

After a small replicator is initially composed and perfected to physically replicate electromagnetic code, the design can be recreated in a larger version. Once the larger version is created to the size of being able to replicate a mobile home other replicator designs can be easily composed to a model size that is more ideal for a home and office. This model will likely be the size of a large cabinet. This size replicator could ideally be used to replicate food and products that are used daily. Having an innovative device like this will likely help all people at all levels of society. I personally believe that after this technology is produced it should be inexpensively distributed to every home on the planet after we can very affordably replicate it again and again.

To accomplish such a task we could also build a larger replicator than the one designed large enough to replicate a full sized mobile home and smaller home replicator model. There would be fewer of these type of replicators than any other model since they would likely be manually assembled in the same general way that all electrical machines are manufactured as I write this book. These model replicators would likely be limited to a number of five or six for each of the majorly inhabited continents on earth, save maybe Australia since it only contains two countries. Each continent could establish a committee composed of representatives from each country to help utilize these vast sized model replicators to replicate the models that can integrate/replicate an object/subject the size of a mobile home for every country within each continent. The continent committees could then put that model replicator in a storage unit until there is a problem with one of the country's replicators and it needs to be replaced. Each country can begin creating the home/office sized replicators for each of their citizens' family's, as this size of replicators will very easily make the life needs to be met. The replicator can in itself be replicated and produced by the in home replicator units to be much smaller to replicate specific electromagnetic signatures of particles, waves, fields, and energy flow.

The Replicator and Future of Our Earth

The replicator will be utilized to reproduce nearly anything for low cost. The replicator will change many current jobs and career positions that will likely be modified after replicators are in place. Most mining jobs of all kinds will no longer be needed, as all minerals, oil, and even diamonds will be electromagnetically replicated. There will no longer be any "black lung" and other terrible diseases or environmental impacts that result from mining or pumping oil. Some factories may eventually close since any product that can be made perfectly will be identically replicated for the cost of electromagnetic energy production, or protons and electrons, so fewer factory workers will be needed. Fewer crop and animal farms will be needed since crops and meat can be replicated. Forests and wildlife will be able to retake thousands of acres of current crop fields that will no longer be needed and effectively be able to create more oxygen in the atmosphere throughout the environment, as all species will be able to prosper.

Hunting and trapping animals for food and furs will no longer be needed as meat and furs can be replicated. Many prescription drugs will be able to be replicated as needed, so outlandish costs for any drug of any kind, to treat any problem, will no longer come at high costs. The majority of insurance companies won't change, since health, home, and auto insurance will still be needed, but insurance expenses may cost much less for treatment, repair, or rebuild. Basically, all of humanity can enjoy a brighter future as we all become able to live easier and more productive lives during the twenty-first century. Real estate will only take a day or two to replicate a building, utilizing a new preprogrammed building machine, as construction workers will do much less individual work. Corporate industries that have obtained too much power over the course of humanity during the twentieth century may, or may not, consistently loose power that was built up during the twentieth century where humanity is empowered, rather than enslaved, by powerful technological innovations.

The smallest specific electromagnetic signature model replicators will likely be commonly utilized to create water molecules at the faucet of our sink and in our bath rather than incoming plumbing. These units also may be utilized to make food, toiletries, and medications, or basically anything that a person will use on a regular basis. These units could also be created to feed any regular addiction with an endless supply. This will make the water treatment plants a thing of the past as water will be as pure as the particles being replicated endlessly or until they are turned off to cease replicating the specific electromagnetic code with which it is encoded. Eventually newspapers may be uploaded to and produced at the news stand or newspaper box where they are commonly purchased while other people would simply replicate internet news on a news style paper on the home model replicator till their heart is content. Of course, there could be just the food, and not product, replicator which may be the best place to start.

The replicator can make the basic life needs of all people of humanity, being food, water, and shelter rights in our lifetimes. Replicators will be utilized to feed the world's population of humanity while using fewer crop fields throughout the world. We will likely be able to replicate oxygen and ozone layers using new replicator technologies to clean up the atmosphere and air we breathe. The technology is currently proven possible, but still looked upon with disbelief and irony. How can we get something from nothing? Everything, including the air you are breathing, is composed of electromagnetic particles, fields, waves, and energy flowing; and interacting with each other and your body. Does "nothing" really exist any longer? We may never know, as that is the nature of irony. With a universe composed of pure positive and negative electromagnetic energy particles, waves, fields, and energy flow our universe and everything found their in is by nature ionic, in addition to being ironic.

<u>Disintegrator</u>

Basics

The disintegrator should most definitely be created to work in conjunction with the replicator, to take care of one of the problems that have existed for humanity since before time was recorded. What do we do with waste? Humanity has collectively been creating waste across the lands and waters of our planet for millennia and we have only collectively recycled small portions of waste on some level since the beginning of production goods for use. We have all reused things or found different uses for items as passive as bones from the game our ancestors hunted, trapped, and killed to eat dinner; as some of the berries in the brush upset the stomach. Human waste has compiled vastly as the population on the earth has continued to increase to consume more of the planet's environmental resources, including every living thing on earth, as our species leaves more waste on the planet than all other species combined. Areas of our planet are reserved just to store our waste, bury it, and wait for it to turn into compost. Some things like styrofoam will take many centuries to simply disintegrate in the top layer of the earth's crust as a result of the earth's electromagnetic core, that in combination with the atmospheric layers causing "gravity" to occur on an electromagnetic level in such waste to separate the subatomic energy particles, waves, fields, and energy flow in natural decomposition. The speed of light squared equals energy divided by mass ($c^2 = E/m$) and mass equals energy divided by the speed of light squared ($m = E/c^2$).

A disintegrator will serve two primary purposes that include reducing all waste and putting electromagnetic energy back into the energy grid after the proton is utilized in conjunction with the electron. Having a disintegrator initially in every community, and eventually in every home, will inevitably put electromagnetic energy back into the grid of every community that utilizes them as their power plants may need to create less energy, in effect cutting down collective pollution to the environment and atmosphere of the only single planet that we have to reside upon within our solar system. We should take care of our planet so it is able to continue to take care of us, our kin, and hopefully many millennia of generations of humanity to come. Future generations can begin to enjoy the cleanest collective existence on our earth, rather than being the dirtiest creature upon our planet since history began, even though we do not smell the worst.

Disintegrators may be able to be utilized to collect engine and power plant exhaust fumes as well during our transition to convert to totally clean energy and utilize electromagnetic vehicles. This innovation will clean up the planet on many levels to help eliminate waste and to reduce the physical labor of work performed each day as you will eventually never have to take out the trash again!

The General Design of Disintegrator Units

A disintegrator will likely have a capacitor added to the initial unit design to also make it easier to operate as energy is to be pumped back into the grid when an object is disintegrated, or to store protons and pump electrons back into the grid. A disintegrator will simply have an electromagnet port hole of sorts that may fit on the top or middle inside of a trash can or maybe just be at the waste disposal yard, which will separate negative and positive electromagnetic energy particles, waves, fields, and energy flow to add the electromagnetic energy into the grid, or simply expel it into "nothing" energy. Disintegrators will likely need to have precautionary measures taken in the design to insure that the technology is not misused in the home.

There will be different sized disintegrators created for various types of waste disposal tasks outside of the home. A disintegrator will likely become a common feature in all of the EM vehicles that I note in a later chapter, so the interiors of vehicles and roadways do not get polluted. There will also model disintegrators that will be utilized for large tasks like demolition. The larger models will be portable and likely come in sections as may be utilized to work together at the four corners of a building structure that has been evacuated to take the building down in moments. There will likely be next to no sound as a building will come down evenly during demolition when it is converted directly into electromagnetic energy evenly as the earth's gravity brings the top of the building closer to the ground.

The Disintegrator and the Future of Humanity

The disintegrator will be an advantageous and powerful part of humanity's future, as it will help us all by converting all types of waste directly from product to the electromagnetic energy particles, waves, fields, and energy flow from which it is composed. By utilizing the disintegrator humanity will be able to turn our garbage landfills that have been collecting for over a century, into pure, clean energy without burning any of it or allowing the trash piles to continue to release methane into our atmosphere as it sits there wasting away. It will be beautiful to live on a planet where there is virtually no waste created by the dominant self aware beings on earth. Unfortunately, the disintegrator may also be used in future intergalactic warfare, but hopefully we can form an EM weapons treaty here on earth to avoid such acts among humanity. With the disintegrator we would be able to recycle anything into pure energy to support society without waste!!

Teletransporter

Basics

As I write this everything and everybody on earth is currently manually hauled from point A to point B over a period of earth time and distance between time zones mostly via internal combustion transportation. A teletransporter is similar to a replicator in the way that it utilizes the electromagnetic composition of all matter to move it from place to place. The teletransporter will essentially break matter down into pure positive and negative electromagnetic energy particles, waves, fields, and energy flow; convert the electromagnetic energy to a different form of electromagnetic energy waves that will likely be able to travel faster than light or be more easily to detect and harness at the place it is being received. Then send the electromagnetic energy of the specific form of matter, receive the electromagnetic energy from the point A from which it has been sent to the point B where it is being sent. Then convert the electromagnetic energy back to pure positive and negative electromagnetic energy particles, waves, fields, and energy flow that is bound in the electromagnetic signature from which it was before it was broken down into the pure energy from which it is composed. Basically teletransportation will be the use of a replicator scan and disintegrator at point A, then the use of the replicator scan and a replicator at point B.

Initial teletransportation experiments will be from point A to point B teletransportation units, but I am confident that we will likely be able to create teletransporters that work in a manner similar to those seen in, "Star Trek", someday after we better understand the electromagnetic spectrum and all of its potentials. Matter will literally be broken down modifying electromagnetic state, sending the electromagnetic energy signature between teletransporters, and then modifying the energy back to initial electromagnetic signature at the destination teletransporter. After electromagnetic positioning systems become more advanced "Star Trek" style teletransporters will be created to focus in on a body or an object's electromagnetic signature frequency, as everything exists as electromagnetic particles, waves, fields, and energy flow. So, it will only take time before that style teletransporter is created to be used on deep space explorative vehicles. To start we can make point B teletransportation units that also act as an environmental mobile exploratory device that can be sent to a planet to sample the atmospheric layers and electromagnetic content of the planet before anyone is teletransported down to the surface of the planet through a unit B.

This technology will change humanity on many levels as it will help to end our addiction to oil, as well as the airplane, train, and automobile. Teletransportation will initially only be utilized on our planet earth. It will not be useful for space exploration until we are able to directly convert electromagnetic waves into bound electromagnetic particles, waves, fields, and energy flow without a point B teletransporter machine with which to do so. Of course product transportation costs will also drop even more. As the replicator will make it so any product design can be downloaded and replicated, the teletransporter makes it so that the cost of moving a product from point A to point B will cost virtually nothing. The teletransporter by itself, and in combination with the replicator, will also change medicine forever as invasive surgeries will no longer be necessary as any abnormal tissues that previously needed to be physically removed can be directly targeted and teletransported out of the human body. We can already easily scan the human body to get an electromagnetic signature that may be used to identify cells or tissues

40

that need to be removed using magnetic resonance imaging and other techniques.

After teletransportation machines are created harmful bacteria, viruses, cancers, and diseased cells will be directly removed from the human body via teletransportation after they are electromagnetically targeted. Just because we can move a human and all their electromagnetically bound cells from point A to point B, does not mean that we have to move all their cells, we could just remove specific cells from their body. Viruses, cancers, and diseased cells are noted to have electromagnetic composition of a certain type to show that they can be electromagnetically targeted to be directly removed from the human body when the teletransporters are created as unified physics has undeniably proven it to be possible!

The General Design of Teletransporter Units

Most basically, the teletransportation units will convert electromagnetic particles, waves, fields, and energy flow to another form of electromagnetic energy after an electromagnetic signature scan, transmit the scan, disintegrate from the point A unit, receive electromagnetic scan that has been transmitted at point B, and then replicate the electromagnetic scan back to the particles, waves, fields, and energy flow as noted in the electromagnetic signature. Or, if it were a "Star Trek" like transporter, the ship's unit would have both point A and B capabilities. Like replicators and disintegrators, teletransporters may also be created in different sizes for various purposes. The majority of teletransportation units will likely be 5 meters tall x 3 meters wide x 4 meters deep and be made to work from unit point A to unit point B. These sized teletransportation units would be the perfect size to teletransport the majority of all people and objects up to the size of a car.

The teletransporter units will likely be relatively simple to create. The first functional unit will be the most difficult, but in a sense we have already solved half of the problem. We are currently able to convert electricity into different frequency of electromagnetic waves to transmit them and receive them. So, do not believe any person who tells you that you can not convert one type of electromagnetic energy into another since humanity has clearly already done just that! Just look at the electromagnetic spectrum and it is evident that we can make, transmit, and receive electromagnetic waves like radio waves. We simply need to expand the same general type of process to transmit converted electromagnetic signatures via electromagnetic wave energy before disintegration, the disintegrate in point A and replicate at point B. Teletransportation will work in a way similar to that of a central broadcast radio that can be utilized together with another unit of the same model to transmit electromagnetic signals from point A to any other point B, while both units are on the same electromagnetic frequency, and back again. Teletransportation units will be able to disintegrate and replicate so they can work going both ways.

The Teletransporter and the Future of Humanity

The teletransporter will hopefully help to end our addiction to fossil fuels to power internal combustion engines currently used to help us to get from point A to point B. To start we could begin to utilize teletransportation units to replace mass transit stops. This could be accomplished relatively simply by installing relatively small bus stop sized terminal teletransportation units that are connected to go directly from the place they reside in the public to a central transfer area where you will be able to walk through another teletransportation unit to get to the area of the city where you which you reside.

Airports, bus, and train stations would be collectively reduced to a central teletransportation stations where an individual could be teletransported to any place on the planet almost instantaneously. Utilizing this type of technology may begin to cause a minor biological time warp of sorts for the human body, especially if repeated exposure to excess amounts of sunlight or darkness which by itself can have negative effects on the human body. Getting to the opposite side of the planet in an instant will likely cause a different type of jet lag.

Relative earth time warps of sort may occur on many levels. A couple days vacation in space every few years in order to be exposed to real "zero gravity" conditions that will in effect help the human body more easily repair all tissues as a result of doing so much less work for motility. This type of improvement to natural biological functions of the human body will likely help humanity to begin living longer and healthier lives on a relatively "smaller" earth with less pollution. By no means will we need to learn how to fly a spaceship to the moon to help our bodies to recover from regular usage of long distance teletransportation units. It may begin to be easier to meet someone than to spend the time to log on and off of video chat online or messaging. Families could live or work half a globe apart, but still all meet at the same place for a dinner. We can all begin to live more interconnected to the rest of the globe, as an individual could someday literally be able to go take a stroll through a different city on the planet every earth day that the person still retains their soul within their body.

We could always simply teletransport food, goods, documents, physical evidence, other objects; and never allow humans to locally over use teletransportation devices, as EM planetary vehicles will be able to transport people just as quickly. Teletransporters will be ideal to utilize as humanity begins to take a vacation to the "life doc" hotel and resort on the moon every five years to improve our physical health in the relatively near future. Artificial gravity will be utilized around twenty one earth hours of each earth day of time in space while we enjoy three hours of weightlessness where our cells do no physical work in one hour stretches. You would enjoy a ten hour evening/morning of artificial gravity where cells physically do less work in outer space, then have an hour of zero gravity before breakfast, then again seven earth hours later, and again seven hours after that. It will be an amazing way to expand the biological lifetime by many years, maybe even an extra lifetime from what is currently considered a "lifetime" on earth.

EM Vehicle Design

Basics

Electromagnetic (EM) vehicles will hopefully help to end humanity's addiction to oil, as well as to help begin intergalactic exploration. The EM vehicles are designed here in there most basic forms to show the essential aspects of the vehicles that will make high speed travel throughout the universe possible, save the electronic navigational system that will be required to help avoid accidents. The EM vehicle designs are basic, but will vary greatly depending upon the speeds at which the vehicle will operate and the places where they will be utilized to travel. The planetary vehicle will be utilized to some extent on earth as a part of the end for our need for the airplane, train, and automobile; as fuel costs now outweigh the productive use of all of them, save the maglev trains. The interplanetary vehicle will be utilized to travel throughout and further explore our solar system. The galactic vehicle will be utilized to travel at speeds many times faster than light to explore solar systems within our Milky Way galaxy for exploration purposes. The intergalactic vehicle will be travel at speeds faster than the galactic vehicle as it will be utilized to explore other galaxies that surround our Milky Way galaxy.

In each section I have outlined the basic structures that need to be created for each type of vehicle that is noted. While each vehicle is similar in many ways, each one of them has very different potentials to travel at increasing speeds for longer explorative journeys. Each one of these vehicles is designed to travel at great speeds while passengers are not affected by "G forces" as they would be in a modern airplane or space ships. Since gravity has been redefined in **"Gravity NOW!!"** it can now be controlled on many different levels which becomes progressively apparent in this chapter, as each exploratory or transportation vehicle design is undeniably valid!

All of these vehicles will likely be saucer shaped, as that general design will likely be most beneficial when traveling at speeds close to and beyond the speed of light (186,000 miles per second). They will all have an exterior layer of a photo-absorbent film combined with an interior hologram layer attached to refract light and created more light energy absorption for the solar energy conversion system. Utilizing this system with more hypersensitive solar absorption it could be used to produce some levels of power at night and in outer space. Of course the power storage system would need to be powerful enough to last a week on earth without any major power intake for the vehicles on various levels, as more energy will be required for each vehicle to function successfully. Each vehicle will also have an onboard computer to help with vehicle controls, speeds, and electromagnetic navigation. All of these vehicles will have replicators for food and goods supply, as well as disintegrators to absorb waste. They will also have onboard teletransporters in order to move things from place to place on the ship to start, as well as from ship to planets being explored.

Planetary vehicle

A planetary vehicle will be utilized to travel within the electromagnetic layers of the planet earth's ozone. This vehicle will be composed of electromagnetically nonconductive material on the inside and an electromagnetically conductive exterior surface. Just creating a vehicle that exerted negative electromagnetic energy within our atmosphere would make it float in the air, but we need to control the vehicle as well. So it will need to be able to exert negative and positive electromagnetic energy from the exterior of the vehicle, as well as form a negatively charged electromagnetic field to surround the vehicle while it is being used.

This vehicle will be saucer in shape as it would likely be able to travel at the speed of light, or just short of the speed of light considering that there will be only one negatively charged electromagnetic field surrounding the vehicle. Controlling the direction and speed of the vehicle will likely not be done with a steering wheel and a gas pedal. This vehicle will be controlled using the various levels of positive and negative electromagnetic energy exerted from the exterior within a negative electromagnetic field. To move forward the front of the vehicle would exert positive electromagnetic energy while the rear of the vehicle would exert negative electromagnetic energy, as well as exerting negative electromagnetic energy from the bottom to levitate. This forward motion is now possible as this all occurs within a negative electromagnetic field layer, as positive and negative electromagnetic energy attract to pull the vehicle forward and negative electromagnetic energy within a negative electromagnetic field repel each other to push the vehicle forward as well as make it hover above the ground. Directions and speeds of the vehicle could now be changed using these methods.

The navigation system would have an electromagnetic global positioning system [GPS] in it to note the position on the planet, as well as its position relative to other planetary vehicles on the planet. A global EM vehicle positioning chip system onboard map will be needed in each vehicle made in an effort to avoid accidents between vehicles. We all need to know were each other are when traveling at speeds close to the speed of light to help avoid accidents. Although, if near accidents do occur, two negatively charged electromagnetic fields, from within which the vehicles operate, will repel each other if they do come to close to each other. We will also have to start adding electromagnetic GPS sensors to the tops of tall buildings to avoid possible accidents. These vehicles will likely travel at relatively low levels, but will likely be able to travel at heights comparable to that of an airplane. With a computer navigation system most of these vehicles will seem like a chauffeur ride as the driver will only needs to program destination instructions with every ride. Roads may become obsolete or at least sustain much less wear and tear as they may be used for these vehicles to travel at relatively lower speeds on a local level. Driving at the speed of light, one would be able to encircle the earth seven and a half times a second; so we really don't need to go that fast.

Interplanetary vehicle

An interplanetary vehicle will be utilized to travel, and further explore our solar system, as it will be able to travel at the speed of light or slightly faster than the speed of light, at a top speed of warp 1. This is the vehicle that we can use to get to the planet of Pluto to sample the planet's ozone and surface, and then come home in one earth day, which is the equivalent of one rotation of the planet, where the vehicle will leave the earth and return before the same time on the next day. We must remember that biological time does not exist in outer space as it does on earth. Every trip made outside the earth's atmosphere into a zero gravity environment makes the cells of our body to do less work literally reversing the aging process which is apparent by astronauts coming back from extended space mission to be literally biologically younger. We do not want to change the way we age by going on a mission to outer space so this biological clock slowing or reversing can be avoided by making the interior and the exterior of the craft able to exert and absorb electromagnetic energy, with an electromagnetically neutral layer of material between them.

To make the interplanetary vehicle travel at the speed of light or slightly faster the vehicle design would only change slightly too likely have the ability to do just that. The vehicle would be larger in size to include enough area to house a crew of five to seven people for a relatively short period of time on earth. With a synthetic gravity within the interplanetary vehicle caused by layers of negative electromagnetic energy that surround it we would likely be able to utilize replicators and disintegrators on the vehicles. We would also be able to utilize a teletransporter that would be able to function from the ship to the earth, but initially we would probably not able to teletransport exploration units from the vehicle to the planet that we are exploring and back again. If we were to send a teletransportation unit down to a planet that we were exploring we would likely have to leave it behind.

The small change to essentially turn an electromagnetic vehicle from a planetary to an interplanetary vehicle is to add one or two more negatively charged electromagnetic energy fields outside of the one that already exists, from within which the vehicle operates.

Higher speeds will likely be attained by increasing the layers of negative electromagnetic energy from one to three in order to create higher levels of electromagnetic attraction and repulsion on the front and rear of the vehicle. The outer layers of the vehicle would also emit higher amounts of energy within the negative electromagnetic fields that are always on during explorative missions. We will also need to modify the electromagnetic navigation system to add an effective electromagnetic map of the solar system which may be attained in advance by new explorative space probes that will be utilized to prepare for such manned missions.

Galactic vehicle

The galactic vehicle will be utilized to explore our Milky Way galaxy, or at the least the nearby solar systems. This vehicle will be able to travel to our closest solar system, which is about 4.5 light years away, and then get back to our planet in a 24 hour earth day at the take off and landing destination. In order to accomplish these goals we will need to create a vehicle that can travel at many times the speed of light. So, we do not just need to travel faster than light we need to travel at multiples of the speed of light, or 186,000 miles

(299,337.984 kilometers) per second multiplied by a power of two, three, or four. That is not just going twice at fast as the speed of light; it is going a lot faster than that! If we could only travel at twice the speed of light it would still take 2.25 earth years to just get to the next solar system, which is a pointless journey if we are trying to explore the galaxy. This vehicle will likely not be hard to create after the planetary vehicle is created, but the galactic vehicle will be larger to accommodate a larger crew.

34,596,000,000 miles (55,676,865,024 kilometers) per second is the speed of light to the second power which will be the speed of warp 2. This speed will be accomplished simply by modifying the interplanetary vehicle design to add a fourth negatively charged electromagnetic field layer outside of the three that have already been created. While this speed looks a great deal faster, in galactic terms it is still too slow as it would still take almost 763 earth days to get to the nearest solar system which is way too much biological time to waste just to get to the nearest solar system. So, we will need to add speeds of warp 3 and warp 4 to enjoy exploring our galaxy. A speed of warp 3 will be the equivalent of 6,434,856,000,000,000 miles (10,355,896,894,464,000 kilometers) per second which is what we will need to reach our closest solar system and get back in less than a day. To truly be able to travel and explore our Milky Way galaxy, which was at one time humanity's entire universe, we will need to make the galactic vehicle able to travel at a speed of warp 4. A speed of warp 4 will be the equivalent of light speed to the fourth power, or 1,196,883,216,000,000,000,000 miles (1,926,196,822,370,304,000,000 kilometers) per second, which will likely be needed to effectively travel and explore our Milky Way galaxy.

As this vehicle will be much larger, be home to some people, and travel at speeds that are many times faster than an interplanetary vehicle it will need to be vastly improved upon. The vehicle's positive and negative electromagnetic energy output/input and storage amounts would certainly need to be increased on all levels. It will not be to hard to reach the speeds of warp 2, warp 3, and warp 4 to create a galactic vehicle to travel and explore our Milky Way galaxy. Each warp speed will be easily reached by adding another negative electromagnetic energy field for each one, after increasing vehicle surface energy output/input. More energy and more negative electromagnetic field layers mean more power and electromagnetic thrust/pull to move the vehicle in a direction at relatively higher speeds. Exploring our galaxy will be fun, but not nearly as enjoyable as exploring the galaxies throughout the universe!

Intergalactic vehicle

An intergalactic vehicle will be a bit bigger than the galactic vehicle, as it will be home to more crew members and be able to travel at speeds of warp 5, warp 6, and warp 7 to enable it to be used to further explore the universe and other galaxies beyond our own Milky Way galaxy. This vehicle will be defined as intergalactic when it is hopefully able to get to and from the closest galaxy and back in an earth 24 hour day from the origin point of take off and landing. This vehicle will travel far beyond the concept of FAST as ever perceived upon earth, but the universe is larger than we can see from our planet using the telescopes orbiting in the layers of the atmosphere!! Going faster means more power from the exterior of the vehicle, as well as extra negatively charged electromagnetic fields with each warp speed increase. The intergalactic vehicle will be able to help humanity explore many different galaxies in the universe, each with billions of stars with solar systems to explore. If we need to travel at a speed of warp 8 or warp 9 to explore further reaches of the universe I am sure we could create those vehicles if they are needed.

Outer Space Exploration

These explorative journeys throughout the universe will not come without established rules of conduct on and off these spacecrafts, as representatives of our planet and species that has realized the full potential of all matter throughout the universe. We may not realize the full vastness of the universe, as we must remember that at one time during the history of humanity it was only the size of our Milky Way galaxy. It may be that another level of macroscopic electromagnetic organization may exist beyond the current "universe" that will simply act similar as a galaxy exists right now once that higher level is realized. One of the most important features of all these vehicles will be the level of controlled gravity on each ship. The electromagnetic basis for gravity on earth is what creates age to occur on a cellular level in biological organisms. We do not want our deep space explorers to come home younger than most of their family when they get back to earth, after what only seemed to be a year on a deep space mission without the correct levels of interior electromagnetic energy, may on earth actually be two years. We should try to utilize many different types of "earth time keeping devices", from the very basic to the very innovative. We will be breaking many current "man made" scientific barriers created as a result of prior scientific ignorance due to misperceptions of the universe and everything contained within it.

A Greater, Better, and Healthier World!

Positive Possibilities

Humanity can begin to progress greatly in a relatively short period of earth time on our planet by utilizing the powerful technologies that are proven to be possible and can likely be created in a relatively short period of time. As humans improve our collective health, nourishment, and prosperity around the planet as a result of the replicator, disintegrator, teletransporter, and the new EM vehicles we can all enjoy many better standards of life. Less work needed to be done for all people around the world as we live better lives on a healthier planet with less pollution where all waste can literally be recycled into the pure electromagnetic energy from which it is composed. We can begin to enjoy new products that may literally operate and be powered by the energy that they consume, like the new age lawn mower and tree or shrub trimmers will be. Humanity will be able to patch the ever growing hole in our electromagnetic ozone layers. We can begin the end our more than a century of being addicted to crude oils and fossil fuels by utilizing more clean energy generating sources (solar, wind, hydroelectric, or geothermal), teletransportation, and planetary vehicles. These methods of transportation over long distances may cause relative time travel by moving from one time zone to another that is hours away in an instant will result in the gain or loss of the time on a 24 hour earth time zone clock. We can also expect to see job growth in the space exploration industry as technologies becomes available.

The replicator will revolutionize work and the concept of retail business forever! People may only buy certain groceries at the market once a month since they will be able replicate them as often as they choose after buying a near perfect specimen to replicate. This may be done with most things that are currently considered retail. The teletransporter will revolutionize travel to the point that in a relatively short period of time the airplane, train, and automobile will probably become obsolete. Shipping could be changed forever, as mail and parcel services won't be as needed when people only need to push a code and send button for automatic teletransportation of goods or replication of items purchased online. A truly global community can be created with the replicator, teletransporter, and planetary vehicle which will all in effect save many species of plants and animals while they lower the need of physical labor around the world.

Begin New Capitalism

Markets will hopefully change forever, as things begin to cost less to produce, and will cost less; maybe eventually; very nearly free and anybody can have access to nearly anything, as long as they make a thumb print credit agreement that they will plan on avoiding confrontation or harming of another individual or group of people as long as they work for a career/job a certain length of years, or more. Humanity will change as our collective needs for food, clothing, and shelter may finally be able to be met on relatively low budget while using fewer crop fields and factory machines around the world to produce the goods. We will utilize replicators to upload the electromagnetic code to create any item. Then we can transport that particular item to another person on the other side of the planet via teletransportation or just send the electromagnetic signature to be replicated, which would be even easier. Some corporations will probably fall within a decade of these new technologies being in place, but that may not necessarily be a bad thing.

There may be fewer jobs for people to do on our planet earth, as new technologies become part of more societies. There will also be many more jobs in the expanding industry of space exploration, as electromagnetic vehicles for exploration eventually become commonplace. tThe initial adjustment that will need to occur with this lack of jobs to be done will be shorter work weeks or less work being done by each worker during the same amount of calendar days as now. More people on staff each working fewer hours likely monitoring multiple computers to insure they are functioning properly while being trained to manually operate all programs in case of program malfunction. We must be sure that the programs that control replicators, disintegrators, teletransporters, and EM vehicles are never fully automated to function and self program, otherwise we may end up with problems noted by current science fiction of computers becoming self aware. We must not forget that we will also have "more food" to work with using fewer crop fields as well; so we can plant more trees and develop more land. Costs and prices should drop with work that needs done by each person in a week. Maybe we can enjoy a three day work week so more people can work the same jobs after the price for many things drops due to ease and cost saving process of replication.

As more automated technologies begin to be utilized to replicate products and food, as well as perform many simple mechanical repairs, life for all of humanity will become easier on multiple levels. We will all be able to spend more time with our loved ones as we can all be happy that no person on the planet will die of starvation ever again while living on earth. Humanity will hopefully be able to live in a more caring, accepting, and better future. All cost of production will be at an all time low as fewer workers, fewer if any raw materials will be needed, and the cost of replicating all of them for a penny each or likely even less! Everything; including us, food, the planet, solar system, and galaxy; have an electromagnetic signature that could be digitally stored and replicated for next to no cost!

Food for All

As replicators become commonplace the majority of grocery stores may become a thing of the past, as they will exist online for downloading some food types to replicate to send around the world, but most people will likely scan and replicate any sample they wish. Local fresh produce markets will be important for fresh fruit and vegetables to eat without replicating occasionally. The problem of hunger in communities around the world should become a thing of the past in a relative short time after replicators are in place. Food will probably become one of the first things that can begin to be distributed via replicator units around the world at no cost to the consumers, as material possessions will cost next to nothing to produce with replicators. Unfortunately, farmers and meat producers services will probably not be needed as much, since home gardens, local produce markets, and replicators will reduce the production needs of their jobs to nothing since food production for all of humanity can be accomplished without their services.

The local garden markets will likely make a come back, as many large crop fields will likely no longer be needed to feed animals for slaughter. Fewer animals will need to be killed each year for food as meat; and as a result fewer crop fields will be needed for feed vegetables, while those that are needed can also be replicated. Nature will be able to take its course as trees and plants spread back through many of the former crop fields to produce more oxygen. Humanity will now be able to expand a bit more as all people will be able to go to sleep with full stomachs and cleaner air. It would be nice to do right now, but when the disintegrator and replicator are working together we could utilize the electromagnetic energy from garbage, biological waste products, or anything else to create foods already having had their electromagnetic signatures recorded on the replicator unit hard drive memory.

Health Care for All

As replicators and teletransporters become commonplace, many doctor and hospital services may no longer be needed. All cancer, virus, and diseased cells will be able to be modified to healthy or directly teletransported out of the human body; so expensive treatments that may or may not work will no longer be utilized since they are not guaranteed to work and simply make certain people or corporations rich. Broken bones, loss of blood; and damages to organs, cells, or tissues can be instantly repaired by utilizing replicators; so long term health care may become a thing of the past. Long term disorders like epilepsy will likely have better treatments available since damaged areas of the nervous system that cause that disorder will be able to be physically repaired, so long term drug/medical treatments may no longer be needed after the disorders are physically cured. All of humanity can likely begin to live much longer lifetimes by allowing our cells a couple of days of much less and no work, every other year of our lives over the age of thirty earth years by taking a trip to the space station to enjoy zero gravity. Our earth astronauts consistently come home from long term outer space missions biologically younger.

Miraculous Cures

Medical science throughout humanity will be able to progress greatly after these new technological innovations are in place. All of humanity has suffered as a result of various diseased cells, viruses, cancers, ALS, MS, and various forms of autism on some level since before history began to be recorded. There have been ailments like the Black Death that plagued much of humanity for generations until the collective population of the planet was negatively effected as a result of all of the death that occurred. Throughout history travelers from far away lands have spread terrible diseases, like smallpox, to foreign areas of our earth to negatively affect other civilizations. As we continuously progress by utilizing new innovative technologies like replicators and teletransporters we will be able to feed and cleanse the human body more easily than ever before throughout the history of humanity. I hope and believe humanity will turn the corner, and the earth will become one of the greatest places to enjoy the miracle of life in the universe, where humanity controls biological aging by utilizing teletransportation to take a couple days vacation on the moon every other year of adult life over the natural age of thirty earth years.

Healing diseases and ailments will become very easy, as we can all take a cue from Jesus and Buddha, and begin to utilize electromagnetic energy to cure others of any negative physical human condition. The mind will still have problems, but likely fewer of them since most basic life needs will be met. Teletransporters can be utilized to directly remove diseased cells, virus cells, cancer cells, cells affected by ALS, and cells affected by MS from the human body so that they no longer negatively affect homeostasis. Replicators can be utilized to reproduce blood cells, plasma, neurons, brain matter, nervous tissue, bones, and all other parts of the human body so that people will never again have to live with the loss of a limb, or simply make sure that no one ever again suffers from low blood sugar due to diabetes after blood sugar replication treatment in the relatively close earth future. Autism and other mental disorders that result from central nervous system damage may soon be able to be cured, or at least much more effectively treated, by reconstructing the damaged neurological tissue. Children will hopefully no longer be forced to grow up with a damaged brain or mental deficiency ever again when these technologies are in place. Suffering has a cure and can end with unified physics technologies!

A unified physics curative cancer treatment process can be done simply by utilizing the powerful innovative technologies of teletransporters and replicators. An electromagnetic scan of the body and the primary area affected by cancer cells will be done to identify the cancer cells to be teletransported out of the body to insure electromagnetic signature of cancer cells to that of normal cells. The cancer cells will then be directly teletransported out of the body, with normal cells of your own DNA being replicated in order to supplement your existing healthy cells if needed. Full homeostasis of the body will be able to be restored with three electromagnetic procedures that could likely all be completed to cure any cancer type in an afternoon, rather than even go to a second cancer treatment appointment. It will be beautiful to heal the human body in such an absolutely waste free process of such a great power of control over all of the electromagnetic particles, waves, fields, and energy flow from which the human body is composed.

Nurses and retirement homes will likely still be needed, but the majority of doctors work will change since some medical and pharmaceutical treatments will no longer be needed when conditions can be cured instantaneously utilizing teletransportation and replication. All of humanity will be able to enjoy a healthier existence and longer lives as a result of these new technologies. The most fun part of transitioning to utilizing these new technologies will be making many hospitals, doctors, and pharmaceutical companies to reducing their prices, not profits, each year, while watching them keep steady control in a lesser inflated economy. All of humanity will be able to enjoy better, longer, healthier lives after the transition to these innovative technologies has been made.

As I write this book I know that many people need drugs to solve and/or deal with their problems. I know first hand as a former pharmacy technician and a man who has lived near three taverns where people enjoy drugs, both prescription and recreational to become comfortably numb to their problems. Or, the individual is helped by the drug to draw on a deeper state of meditation and consciousness in which a solution to their problem is realized through greater visionary understanding of the problem. Then they may physically accomplish the goal at a later point, in addition to then be able to focus their meditative energy on their actual feeling of accomplishment and deeper state of personal enlightenment.

The End of the Drug Wars

As replicators and teletransporters become commonplace the cold "war on drugs" will be truly realized as inevitable lost cause! We must all realize that the current drug wars are more about government power and control of the private lives of citizens, rather than for the greater good of the population. People will be able to find the best sample of their favorite pharmaceutical or recreational drugs and replicate them to their hearts content. The only thing that societies of the world could do would be to prohibit drug use in public places that did not include personal transport vehicles or homes. If somebody walks down the street on something, oh well, unless they choose to do something stupid against another person or group of people. Recreational drug usage has risen since the "drug wars" began to advertise how to use drugs and a relatively steady level of the population have been using recreational drugs since then.

The drug wars account for more than half of the United States prison population. Recreational drugs being illegal and the "under the table" money that can be made are also one of the primary motivators for crooked law enforcement officials. In the United States the drug war/prohibition started as a result of very clearly bias reasoning and the loss of federal funding of alcohol prohibition that had just been repealed. When drug laws were started they got around them being unconstitutional by creating necessary tax stamps for marijuana that were never printed. The bottom line is that the United States recreational drug laws are based on the fact that Harry J. Anslinger did not want to loose his paycheck after prohibition was repealed!

It can easily be interpreted to acknowledge that I have the inalienable right to use any drug that I so choose in the privacy of my own home in the United States of America! The Declaration of Independence, which is at the heart of the first US Constitution, since it was written many years before to form a foundation on which the United States Government Laws and Ideals are built. Drug laws are clearly in violation of United States Constitutionally Guaranteed

Rights of Amendments 1, 9, 14, and 21. The first amendment guarantees freedom of religion and insures no law prohibiting the free exercise thereof, so all people who are shamans or enjoy utilizing "drugs" to gain a deeper state of meditation and connection to a higher power already have had drug laws made to inhibit their religious practices which is unconstitutional; and one reason that Anslinger used tax stamps that were never printed to keep his paycheck, but since then unconstitutional state laws have been made. The ninth amendment is in place to keep rights of the people noted in the constitution from being denied or disparaged, but they have been. The fourteenth amendment notes that no state shall deprive a person of liberty, which is the right to think and control ones own actions, and the drug laws clearly are in violation of liberty of people to choose for themselves without harming others. The twenty first amendment is the repeal of prohibition of liquors which should have put Anslinger out of the job. History shows that he blatantly used the Mexican immigrant workers that helped our country during World War II as scapegoats for relaxing through the use of marijuana, in the privacy of their own homes where law has no warrant, after their hard days work.

I clearly noticed that historically our country did not begin the "drug wars", starting with abolition, until after every living person in the USA had no living relative who knew somebody who was alive for the Boston Tea Party. There are many herbs that are legal that are much more potent than any of the drugs made illegal by state laws. It is also not surprising that the eastern USA tends to have stricter recreational drug laws than the western USA, having an ocean of land in between them with mixed standards. The current "tea party" in the United States is, every day, acting more like England before the American colonists declared their independence during the great Revolutionary War then founded the United States of America! I, as a United States Citizen have the Inalienable Rights of Life, Liberty, and the Pursuit of Happiness, the latter two of which have been getting trampled on for years. Underworld money continues to change hands; while various drugs continue to be produced in, or come into the USA.

A Peaceful Earth Population

Once humanity has created replicators we will begin producing more replicators of a reduced size for homes and small communities of all people around the world. This is a blessing for humanity. Curative medicine will be reduced to a set of electromagnetic scans. The first scan will be to diagnose the abnormal tissue, disease, body malfunction, cancer, bacteria, etc. The second scan will be to correct or remove the diagnosed problem. A third scan of sorts will likely be done to insure that homeostasis has been restored. Medical science will change forever, as humanity will be able to utilize natural herbal medicines to remedy a cold, cough, or headache, and medicines at the drug store, or buy one package of medicine, scan it, and replicate it whenever you feel the urge to do so.

The logging industry will likely fade quickly, as lumber will easily be replicated, so trees will no longer to be cut down to make supplies; only for construction and aesthetics. Soon homes will be programmed into an electromagnetic assembler and the materials will be replicated already assembled, so architects will still be in business, but some home builders may have a change of position. The environment will flourish as we transition to even cleaner energy production and end humanity's addiction to overpriced oil for transportation and industry, as the internal combustion engine will eventually become obsolete as the primary source of transportation.

Everything will be able to be electromagnetically recycled with disintegrators, so even nuclear waste and landfills can be disposed of properly. Food, clothing, and shelter will likely become a right rather than a privilege as replicators become commonplace and everything can be made at the best quality with the absolute lowest cost. Retail business will change forever as many locations, like malls, may no longer be needed when everything can be replicated and teletransported.

Most businesses will still exist; although factory, production, farming, produce, crop, and many retail business industries may decline a bit as new technologies become commonplace. Service positions like teacher, professor, lawn care, nurse, pest control, fire department, emergency medical technician, law enforcement officials, politicians, and entertainment industry professions will all be relatively secure in their jobs. Professions may simply become something to do, a passion, a talent, or a way to define ourselves. People will hopefully be able to work shorter hours for a shorter period of their lives at a position they hopefully enjoy, while people tend to retire younger, at around the age of fifty earth years. Humanity will collectively be able to spend more time with family and friends.

Work will no longer rule the people as we will be able to enjoy the company of our friends and families more often, but as citizens of humanity on our planet earth we cannot let ourselves get lazy either, as we should all be sure to regularly physically and mentally exercise as well.

We should begin saving people money by reducing automobile pollution and the need for the oil industry quickly by distributing replicators to gas stations around the world so the cost of gas will drop to next to nothing since gas stations will never have to buy fuels to fill up their storage tanks ever again. To complement such a step towards the progress of humanity we would also need to create specialty disintegrators that would be affixed to the automobile exhaust pipe to turn fuel exhaust fumes directly into electromagnetic energy or convert particles of methane to oxygen, rather than allowing the internal combustion engine vehicles of transportation to continue polluting our earth's atmosphere. On the other side of that coin specialized replicator EM vehicles will likely also be created to literally hover at certain heights within the earth's atmosphere to begin rebuilding the layers that have been vastly depleted over the last century since the automobile and electricity production have become popular innovations to make life on our planet earth easier, just as the horse and candle were phased out over time. Oil will eventually never need to be pumped from wells or transported across the land and sea of the earth as fuel ever again, and the natural oil spills of sorts will be cleaned up relatively easily by utilizing specially made electromagnetic disintegrators.

Fast food jobs may also be one of the first major corporate capitalistic industries to quickly drop like a rock after the replicator is in place; or they may skyrocket if the fast food is made perfect to be replicated! First they will begin replicating food at the burger shoot from the kitchen, but who will go and buy a burger from a fast food place ever again after a person scans the last one they bought and can now replicate it! People will be able to share food recipes with an on hand sample of the finished product to send it via teletransportation, or the electromagnetic signature code from the scan of the finished product the last time that it was made from scratch can be emailed as an attachment to be replicated later. Did you forget to scan your favorite burrito before the restaurant closed down due to lack of business interest? Well somebody will likely have scanned that style burrito freshly made and posted the electromagnetic signature code online, so we should all try to share food electromagnetic signatures online when we can. Then we could begin to download and replicate your food, or make your best of everything and scan the meals of perfection one by one to store the electromagnetic signatures, and then never have to cook that same meal again if we choose not since we could just select the electromagnetic code and replicate it at a later point. We may be able to put an E-Block code in the food wrappers so food could not be replicated before unwrapped, or something like that. Of course, there may be a catch that restaurant food must be fresh made, or could have a catch like it can only be replicated 2 times after purchase.

The value of money will be able to increase like a rocket and international monetary debts will hopefully eventually be more easily negotiated as a part of humanity's period of monetary adjustment on the evolutionary chart. Physical goods, precious metals, and monetary credits of sorts have been traded between people in exchange for goods and services for millennia. I hope and expect electromagnetic energy production to become environmentally friendly on all levels. Soon all power plants will be able to be outfitted with disintegrators to help them effectively recycle all of their waste directly into electromagnetic energy particles, waves, fields, and energy flow that can also be utilized to increase electromagnetic power production output. Humanity will be able to progress to a greater level of conscious physical existence within the now much larger universe that exists outside of our earth's solar system as it is awaiting human exploration in Galactic and Intergalactic EM vehicles that will be composed relatively soon.

Possible Technology Backlash

Corporations and some government agencies around the world may initially try to stop the inevitable money gaining value from occurring as everything (food, clothing, shelters, etc.) will be able to be produced for pennies or less as replicators and disintegrators are utilized. Jobs may be lost around the world as corporate sweatshops become pointless entities since goods will be produced for less using replicators. The same people working those jobs could be more easily be fed using the same technology. The "eight hour" work day may need to be adjusted after less work may physically need to be done to make it so more workers each work less time to accomplish the same amount of production. That is until we start exploring outer space more, then jobs will soar, starting with a new ships and then a vacation spot on the moon. Profits may be reduced as technologies will change forever. Elections in my own United States will no longer be able to be indirectly bought by corporate interests who help to get ignorant voters to believe what they want them to believe through media like television and newspapers that print or show what they are told whether it is actually true or not. Most voters in the USA can be duped into believing anything which has not been proven beyond doubt as the citizens of the United States.

It is sad that the United States of America and its citizens have clearly become socialized since that 2000 "election." As soon as our government "bailed out" corporations and businesses we became socialized since at that moment the tax payers of our country and our "elected" government began to financially support businesses and true capitalism ceased to exist. Capitalism is about business succeeding and failing as a result of their owners decisions, without safety nets, but unfortunately now the US Government is a safety net. The Hostess snack food company just declared bankruptcy, which if happened twenty years ago would mean they are going out of business, but now they will just refinance thanks to the US

Government lending tax payer money or money that our government borrowed. That is socialism at work!

Humanity, but most definitely the USA, has become socialized on another corporate level that is also making our law enforcement officials lazier and slower. As more cameras go up around our towns at street corners and street lights our privacy is evaporating. Of course the real socialization to which I am referring is the global positioning system chip that the majority of US citizens are carrying around in their pockets everyday. A GPS chip was used during my childhood, just before the end of the cold war, by spies to track people after the chip was planted on the target individual or vehicle. Now anybody with a cellular phone can be tracked by anyone with an internet connection from anywhere at anytime. Privacy has become an illusion and only socialized people allow themselves to be tracked. I have found few people who have used their GPS cell phone to find their way to a new place without a map in hand more than once.

Governments and corporations of the world will keep their power, but not that currently comes from monetary wealth. Collectively both entities commonly buy silence and create falsified evidence through agencies put in place to protect citizens from corporate and government predators. Bribes will hopefully no longer be able to help such people to buy their way out of problems and consequences of their negative actions. The replicator, disintegrator, and teletransporter can empower humanity on many levels as we end our addiction to fossil fuels and other multibillion dollar industries. If people so choose they will be able to take home their favorite meal from a restaurant, scan it, and replicate that meal identically as many times as they wish. The people of our planet earth may again need to have many revolutions against the corporate industries with current monetary power that they will inevitably use to try to suppress these new technological breakthroughs that may put the majority of them out of power. Government officials of the world that do not support making the lives of all the people of our earth simpler, easier, and better through the use of innovative technologies, since they support big businesses, will hopefully be replaced by government officials that accept and embrace new technologies that will help the planet, as well as everything that lives upon it. The world can become a better place to enjoy our existence while we end our fossil fuel addictions, clean the atmosphere, enjoy cleaner air, and enjoy the progress of all of humanity on many levels.

Outer Space Exploration Standards

As humans begin to explore our galaxy and other galaxies we will need to insure the integrity of our species throughout the universe by creating intergalactic exploration standards. We should be prepared to take part in an intergalactic senate, or some such political organization of solar systems, as predicted in current day, "Star Trek" and "Star Wars" works of science fiction. We should never make unnecessary contact with any alien species; especially those species that are self conscious or dominant on their own planets, but still unable to accomplish intergalactic travel since they still do not understand that the universe and everything within it is composed of electromagnetic particles, waves, fields, and energy flow. Those types of self conscious beings, for which we are all a part as I write this book, are not truly prepared for the technological breakthroughs and great peace for their dominant beings through the use of electromagnetic technologies or intergalactic politics. We must be able to expand our understanding of beings that are self conscious and interpret unified physics in their own way to have the technology to travel the universe with a very wide open mind. Cultures will likely vary incredibly throughout the universe!!

Humans are a species that has been destroying our planet earth at increasing rates for the last two earth centuries. Humanity has also not truly realized unified physics to its greatest advantage for millennia, until now, as we have only been conscious of many parts from many different points of view, but never together to be realized to obtain great powers. Our Higher Powers have been defined and redefined on some basic level of pure positive energy/entity/power/emotion/good and of pure negative energy/entity/power/emotion/bad. We can now exist on another level since we now understand the true power of the electromagnetically bound universe and as is everything found therein is ultimately composed of nothing more than varying levels of pure positive and pure negative electromagnetic energy particles, waves, fields, and energy flow. We have all felt that energy running through our body at some point in our lives.

Outer space explorers must be prepared for anything! What I mean by anything is that an intelligent entity of the universe may have very different customs in their solar system. For example, rather than shaking hands, they may show peace by placing their biological excrement tube up to your face and letting it rip! Alien excrement in your face may be an extreme example, but anything could happen. Most likely we will not initiate first contact with alien life forms from outside of our solar system. It will likely happen after we break the light speed barrier beyond a certain point, probably with the EM galactic vehicle. Humanity will be a part of a much larger universe and reality after first contact with other intelligent beings from another solar system. Our home planet earth dominant species of homosapiens will hopefully find ourselves closer to world peace after our suffering has ended with every hungry mouth fed and every cold body clothed through the use of new technologies, as well as a healthier people around the world.

Outer space exploration standards should be basic and simple. We should not interact with any societies that are not able to enjoy galactic space travel. Those societies on other planets are not mentally advanced enough to understand the true conception and potential of the universe as well as everything existing within the universe. We should be sure to send explorative probes to any planet where life may exist to make indirect first contact or try to avoid unnecessary first contact. This will mark the beginning of humanity's existence in a much larger universe that contains other intelligent, self conscious, problem solving beings on other planets, in other solar systems, and in many other galaxies.

EM Weapons

Basics

It is truly unbelievable how many powerful weapons can be created to make the physical existence of any figment of electromagnetically bound matter cease to physically exist or electromagnetically modify the target in some way. The bottom line is that "fire" arms do not work in space since there is not enough oxygen in the vacuum of space to create a spark or for black powder to burn. Electromagnetic weapons will be important to protect our EM vehicles during deep space exploration from asteroids and space debris. They may also be important if an intelligent, technologically knowledgeable, hostile alien species traveling space make their ship and existence known to us after they observe our ability to travel faster than the speed of light. While we must try to keep peace between other intelligent beings of the universe we must also be prepared to defend our explorative vehicles and the passengers on board. We should also never allow our electromagnetically created robot or biological clone soldiers any civil rights as they will literally be created for the sole purpose of fighting to defend and possibly be destroyed while defending humanity and the areas which they inhabit.

Many of these weapons could be used on our planet earth between cultures of humanity. I truly wish to avoid interpersonal violence utilizing these new technologies or another "Cold War" from beginning, which is why I am trying to make the most dangerous weapons that will be created a part of public knowledge. I hope to prevent these weapons from being used amongst humanity and only be created so that they are available to be utilized to defend our planet earth against possible, but unlikely alien invasion, and to protect our deep space explorative vehicles. I hope to see a World Electromagnetic Weapons Treaty between all countries on our planet earth to prevent their misuse. Although, some will be very beneficial for law enforcement officials of the planet to use effect nonlethal weapons to stop criminals at relatively long distances for which a firearm is currently utilized.

Force-Field and Cloaking Device

This technology will be important for EM vehicles while they are traveling around our planet and throughout space. Electromagnetic force-fields will actually be relatively simple to control on EM vehicles. Simply enough negatively charged electromagnetic force-fields are a literal part of these types of vehicles that not only protects the vehicles and passengers, but is literally part of the system that enables motion. The negative force-fields will be in place while these vehicles are turned on to work in combination with the positive and negative electromagnetic charges that are emitted from them. Force-fields will become very important while traveling in deep space exploration vehicles in areas of the universe that are essentially uncharted. We could come out of hyperspace in an asteroid field, but our ship would not be effected by them while in hyperspace, as they would be repelled by the negatively charged electromagnetic force-fields, which will likely have more uses during space travel.

Deep space exploration vehicles should be equipped with a cloaking device to keep from revealing our existence to another self conscious, intelligent, dominant species in other solar systems. The use of this type of stealth during our missions of space exploration may also be important to avoid conflict between other intelligent beings exploring the universe. A cloaking device will simply modify the electromagnetic consistency to reflect the opposing side of the outer electromagnetic shell layer to match the consistence of the space surrounding it to effectively disappear from visual site, as the electromagnetic shields protect the EM vehicle from direct impact with anything else that exists in space moving at any relative speed. Most basically each side of the vehicle's surrounding field will reflect the view from the opposing sides to make it appear invisible.

LASER and Particle Beam Technology

Light Amplification by Stimulated Emission of Radiation (LASER) technology has already been created and a similar device will be utilized in the future for many different reasons in the future of weapons and tools of deep space exploration. A future "laser" of sorts will be able to be produced more easily and efficiently than any of the lasers currently available today. Future lasers will be produced as a result of focused energy on varying levels of electromagnetic wavelengths to become ideally effective in the way that the energy they produce is transmitted toward the object of target. Lasers of sorts will likely be primarily utilized as weapons, as they will seldom be needed to aid work or medicine after we begin to utilize teletransporter and replicator technology.

Particle beam technologies have been created, but can be vastly improved upon for various purposes during space exploration and on earth. The particle beams may be utilized during galactic and intergalactic exploration to attract and repel objects to and from the ship. This may become an important to avoid space debris and possibly to immobilize objects in space by making them cease to move in a certain direction through the solar system or galaxy in which the object exists. We could possibly utilize a particle beam to stop Halie's comet to examine its electromagnetic composition before putting it back into orbit through our Milky Way galaxy. Particle beams will be composed of pure focused positive and negative electromagnetic energy and in order to be utilized effectively the weapon will be charged to work on a certain electromagnetic frequency after a scan of the target's electromagnetic frequency has been obtained.

"Phaser" or Laser Gun

The "phaser", or laser gun, will be a powerful electromagnetic weapon of the future that has been foreseen in many science fiction stories. The phaser will be a weapon that will be able to be adjusted for damage to be inflicted upon the target. This weapon may be utilize to stun, similar to a stun gun, to knock people unconscious, but it can also be deadly on many levels. It will also be able to be adjusted to kill someone and spare the body or totally disperse the energy contained within the electromagnetic particles from which the human body is composed. This will in effect eliminate any evidence that a body ever existed, aside from personal effects at a residence used before disintegration.

The final effect of the phaser is obviously the most dangerous, as murder could occur with no true evidence of persons ever truly dying, except a possible electromagnetic signature. Of course, here on earth we could make the laser guns only able to stun mildly and turned up to stun very harshly, but not necessarily lethal. Big businesses and governments may try to utilize such weapons against private citizens who have established an offensive approach to solving their problems with government policies, as I agree with the citizens that corporations have too much control of my United States Government and State Government. As I have noted, and these USA citizens realize, our democratic United States of America has become socialized to support big businesses and corporations assisting them to feed off of the United States Taxpayers like leaches. It must end eventually, and I hope it does not take the use of phasers to eliminate elitists who try to get in the way of common people from living better lives at a better price, since it would not get in the way of their personal and corporate profits, as the dollar will be worth more! Phasers may be restricted to outer space travel since they can be so lethal, but they may also be made as a nonlethal weapon that can effectively stop another person from a distance on earth or on an outer space journey.

E-Bomb

This bomb will hopefully never be used on the planet earth and would likely not be utilized during conflicts between ships in the vacuum of space, but it may be useful if we ever must invade another planet to destroy their existence. I hope and pray that such a device will never be utilized for any reason, but the possibility of it coming to be created does exist. The atomic H-bomb is one example of an E-Bomb, as all nuclear weapons are another level of electromagnetic energy at work, but are still not the most devastating types of E-bombs that could possibly be created. The type of E-bombs to which I am referring would contain and release pure positive or negative energy in various amounts that could literally disintegrate an entire area of land, not even leaving ash or a mushroom cloud behind, but would probably leave a large hole through many layers of atmosphere above the detonation zone and a few layers of the earth's crust missing from below.

This bomb may have a pressure trigger to it, but more likely it will have a timer or switch to release the pure electromagnetic energy to likely desecrate a planet for generations to come. I hope to never witness such a weapon used for any reason. A person who would ever utilize such a device against any other person, friend or enemy on our planet earth would be betraying their personal existence, their family name, all future generations of humanity from any place on earth, their personal higher power, and everything that they have ever believed in for any reason! While the E-bomb could literally destroy a planet and every living thing upon it slowly after such a detonation; there could be much worse upon our horizon. At some point we could also be forced to betray ourselves, our species, our planet, and all of humanity using literal electromagnetic mind control technology.

Mind Control

This will be one of the most terrible devices that will have the potential to exist utilizing unified physics. The central nervous system is where the conscious and subconscious mind controls the human body. The conscious mind controls bodily movements, actions, and cognition electromagnetically through the central and peripheral nervous system while awake. The subconscious mind controls internal organ functions, as well as memories, dreams, and movements while asleep. The central nervous system makes up the electromagnetic "battery" of the human body as it clearly can be tested using Magnetic Resonance Imaging, CT scan, and EEG brain wave tests. The electromagnetic particles, mostly neurons, which make up our central nervous system, can be modified relatively easily utilizing unified physics technologies.

A mind control device would exist totally independently of a replicator or teletransportation device, as it would function on a range of levels for various purposes. This device could literally be created to tune in to the particular wave frequency on which an individual's brain waves are functioning at that time to listen in to another person's thoughts. A mind control device could also be able to make another person to be put to sleep instantaneously. Unfortunately, it could also be utilized to erase and reprogram memories after a sample of thoughts and brain waves has been confirmed identically after another sample. This could be terrible on many levels. Although, this ability could prove to by very useful during deep space exploration if we do take samples of self conscious beings that clearly dominate life planets in other solar systems, but have not realized unified physics. We would not want to interfere with the progress of other intelligent beings, so if we had to take a sample, we could return it with memory of the event erased from their conscious mind.

Replicators as Weapons

Electromagnetic clones of anything created as a result of design or by sample of any life form that could be made to fight for our planet so human lives are spared. Replicators could also be utilized to reproduce weapons of all kinds. We could literally reproduce clones of electromagnetically designed beings with preprogrammed phasers with one thing on their mind, kill a specific preprogrammed enemy. Although, most of us have been subjected to science fiction stories of such types of beings, whether biological or mechanical in nature may at one point see humanity as a target then try to eliminate us as we some how become the enemy of our custom made protectors. That may likely be avoided by simply not ever taking humans completely out of the loop of goods production, even if it is as simple as turning on and off manually is the only real control. If we allow synthetic brains and problem solving computers or software to do too much of our thinking for us we are dead already.

Being a part of the video game generation may pay off in the long run, as mechanical robot clones could be physically controlled on some level, while biological clones could be literally preprogrammed to think a certain way with certain goals or kill targets, but being biological they would also be created with some level free will. There may be no best way to defend our planet and explorative ships utilizing replicators, but I believe robot clones would be the way to go over free thinking biological clones. Robots would be easier to control, as long as humans never fully turn over robot design and creation to computers or synthetic brains. Although, we may also be able to utilize massive mind control devices in order to control groups of biological clones and other biological clones that we may be defending against. Either way on some level it may simply be like hacking a computer to control you opponent's troops to, in effect, turn against your opponent. It would be a whole new style of interstellar warfare on many levels.

Teletransporters as Weapons

Teletransporters may be utilized to send replicators to planets or the ships of our enemies to share war the old fashioned way with the twist that no humans would be killed during such interstellar types of battle. This means that all supplies could be digitally shipped with armed forces and an infinite number of them be produced for fighting forces on site. Teletransporters will reduce the supply load of all armed forces, but clones do not need to be treated medically or repaired as more could be produced in less time than it would take for repairs or treatment. Teletransporters will revolutionize warfare by allowing the transport of anything in an instant over relatively long distances on earth, but relatively short distances in outer space. We can always disintegrate clone pieces, as well as utilize their energy to create other things or more functioning clones if they are needed.

A Hopeful Future

World Electromagnetic Weapons Treaty

I hope and pray that all of humanity on our planet earth in every country can peacefully become prepared for a greater future as we can reduce the need for major crop fields, end the need for destroying wildlife or forests, lower the worry of earth overpopulation, and end all fear of a horrible, painful death by most virus, disease, or cancer. While we will be able to stop other organisms from killing us we can hopefully vow to stop killing each other so that together we can help the humanity to live together in harmony on earth to progress throughout the universe. Electromagnetic weapons, aside from low level stun phasers, should be utilized only during missions outside of our solar system to defend explorative EM vehicles and their crew. We should try to avoid testing electromagnetic weapons on our planet or our moon to avoid unnecessary destruction of our planet earth and to prevent their use anywhere near our solar system. We should certainly visit other solar systems nearby before worrying about how we can disintegrate a planet and everything upon it in an instant. We should love all of our neighbors on our relatively small, life filled planet earth to try to keep the peace as all of humanity's life needs will be able to be produced with a more valuable dollar and distributed for low costs by utilizing new these innovative technologies are now proven to be possible.

Progressing With More Valuable Monetary Currency

I have heard that money is the root of all evil. That statement may very well be true. Money is sometimes be able to buy all levels of politics and law enforcement officials to allow the distribution of drugs, murder, and other current "crimes against humanity", as they unfortunately turn their heads the other way while they enjoy their extravagant gifts or offshore monetary accounts. Money can become overcome by being replaced by bigger, better, simpler, and more respectable goals. Having the primary goal in life of obtaining monetary currency in order to exchange it for goods and services in order to provide for yourself and your family is still a worthy life path to which the majority of humanity can still enjoy! We may enjoy shorter work days at some point.

Monetary currency has been a part of humanity and the majority of societies around the world on some level varying beyond barter for millennia. It is not often when cultures see a deflation as would likely occur with the implementation of unified physics technologies. Money will be worth more!! To imagine an earth with fewer poor people with a more powerful dollar will be nice. Each person must be able to try to work a given profession for at least twenty five years of their human life, or more; unless disable; and teach two people how to work that profession in the future. There will always be people that wish to change their career paths, but that could be avoided by allowing younger labor and educations to progress to the age of twenty five. Each person could be given a greater chance to find themselves and their true calling to follow it and as a result live a happier, more prosperous life with more personal time to meditate and enjoy teaching the future generations of humanity better life skills.

A Prosperous Life for Every Soul

Every living thing in our galaxy, including our planet, has a "soul" literally defined as the electromagnetic energy that flows throughout every system within all living things that keep them alive on some level. People, plants, animals, bacteria, viruses, our planetary system, sun, and other "living" things throughout the universe all have a soul that flows throughout their existence, from beginning to end. Our souls or the souls of humanity will be able to exist for longer periods of time during their life on earth now that time has been redefined. Biological time for a living organism can be extended and literally be reduced by taking a trip to outer space beyond the earth's outer atmosphere to enjoy a few earth hours every other year experiencing "zero gravity" where cells do much less work than when within the earth's atmosphere. It works for astronauts on every mission to date! We will create rejuvenation colonies on the moon where people can enjoy their experiences in some areas while also appreciating relative earth gravity in others.

We can allow the earth's wild life to prosper and flow across our planet as all life on earth enjoys cleaner environments on land and in the water. The replicator will reduce the need for death on all levels as the creation as lumber, food, clothing, and shelter will all be produced without the need for as many crop fields, slaughter houses, sweatshops, or the current amount of manpower and mechanics utilized to build structures in which to reside. The teletransporter will also enable humanity to get great daily exercise by doing some walking, as well as being able to work anywhere on the planet earth by going from point A to point B in an instant. Humanity can hopefully begin to live in a greater harmony between all of the cultures of the planet as we begin to interact more freely on a day to day basis. The planetary vehicle will literally be able to help us interact in person almost as easily as we will be able to interact online, since speaking to another, in person, is usually easier than typing.

Free Exchange of Ideas and Technology

The internet has been in place and connecting communities and individuals around the world for more than twenty years now. It has, in essence, prepared humanity for the changes that are about to take place as a result of the new technologies that will become available through the use of unified physics. I am certain that we will be able to create verbal electronic translators that will enhance our abilities to better interact with each other on an intercultural basis to help the planet and every living thing on the earth to prosper. We can each help one another to become wiser in our ways, as a truly wise man always asks the question that has not yet been answered. If granted an answer by a professor we can only become wiser as a result. A closed mind is like a morning glory flower that looks wide open in the morning and stays closed during its day in the sun, as well as the night only to shrivel and die.

I hope and pray that humanity can collectively gain the power to accept each other as we exist, rather than trying to mold each person into something they are not. We can all think for ourselves freely without anyone telling us what to think, even though many people prefer to be told what to do and think. We should try to collectively adjust to these new technologies and a larger universe together as both are greater than only one. The many life forms on the earth can now take pride in the fact that on the most basic level we are all composed of the same things, pure positive and negative electromagnetic energy particles, waves, fields, and energy flow which is evident on many physical and social levels.

A Relative World Peace Treaty

This document will initially be in place for only two primary purposes. The first purpose of this world peace treaty will be the free exchange of ideas and information between all nations and peoples of the world. The second purpose will be for all of the countries of the planet earth to be able to drop all interpersonal conflicts on an individual or social basis to defend our planet earth from any alien forces that are clearly hostile towards our planet and the living things that live upon it. This treaty would hopefully be able to drawn up and agreed upon by the governments of the world within ten years of the global publication of the "Power of Unified Physics." Humanity should be prepared to defend our planet after we have created many of the weapons that can be made utilizing unified physics. Although, I truly hope that a minimal peace treaty between the societies of our home planet earth will lead to a greater peace between all of the life that exists on our beautiful planet earth so that we can set an intergalactic example by having our own world peace between life forms on all levels of existence to some degree. This would hopefully be composed by the United Nations in an effort to include all nations of our Great Planet Earth!

Intergalactic Social Relations

Within my lifetime, or within a hundred lifetimes, our humanity exploring a universe so vast will find technologically advanced intelligent social beings who on some level understand the electromagnetic basis of the vast universe and everything within it. In our Milky Way galaxy alone there are more stars like our sun, each having a solar system, than there are grains of sand on our planet earth's surface as well as on our ocean floors. I highly doubt that we are the only intelligent, self-conscious, beings in the universe that has as many galaxies; each with so many solar systems. We should be prepared in many ways for intergalactic politics that will exist in an effort to keep the peace between the various intelligent, social, communicating beings of the universe at some point during the future existence of humanity.

It will likely become a pleasurable experience to interact with other intelligent beings of the universe that we could share thoughts, theories, and ideas about exploration of our vast universe for which we cannot seen the end. There is always intergalactic gossip of sorts once a line of communication has been established to hear of their species' journeys around the universe. We must be prepared for anything! I mean that it will take a group of individuals with vast amounts of patience and self confidence some time to learn how to properly interact with intelligent alien beings that wish to socially interact with our planet's population. For all we know a species of another planet could show their initial peace with another intelligent being of the universe by excreting their bodily waste upon the ground, picking it up, and........ You fill in the blank, as it could be any action to show their good faith. So, we must try to keep peace between other intelligent beings that reside in our universe to avoid any bad blood.

Remember that they could have much knowledge of the powers of the vast universe that we do not yet understand. Throughout world medicine, our earth's humanity collectively believed that the human body was born with all of the neurons that it will ever have throughout the life span of each person. This statement is clearly untrue as the human body skeleton, and nervous system clearly grow throughout the lifetime of a human being, or at least until the end of puberty. During the first decade of this millennium science finally found the area of the nervous system that created neurons. Things are not always as they seem since senses of the body and perceptions of the mind can be deceived on many various levels.

Progress of Humanity

The self realization of our existence on a spherical planet in a vast universe that is much larger than our Milky Way galaxy by itself is relatively fresh knowledge that has become evident to humanity during a very short period of existence within the life span of our planet earth. The collective unconscious holds the future noted through fiction, which commonly predicts the future, as well as through a deeper understanding of our history by observing the varying understandings of the pure positive and negative electromagnetic energy particles, waves, fields, and energy flow from which everything throughout the universe is composed. All people on the planet earth will become able to enjoy shorter work days, possibly shorter career lengths before retirement, healthier existence, longer earth lifetimes, and a teletransported trip to one of our earth life revitalization colonies on the moon where people will physically become younger on a cellular level, and hopefully become mentally rejuvenated, as a result, to then becoming able to live longer more enjoyable and prosperous lives. We all will hopefully start to be able to play physical sports with our twenty five earth-year old great grandchildren while feeling great rather than tired, rundown, and foggy minded.

We can begin to utilize clean energy around the planet on many different levels, while we will also hopefully be able to build a machine that will effectively repair the electromagnetic layers of our earth's ozone. The planet and all of the life to which it is home can begin to enjoy a greater clean healthier place to live where anything can literally be electromagnetically recycled to produce energy, as a variety of trees at the age of thirty years and wild forest are replicated to replace the fewer crop fields that are no longer needed to feed the populations of the earth. We can all thank our higher powers for making a "heaven" on earth possible, as all life will be able to prosper shortly. Manually controlled robots equipped with design software and replicator abilities to electromagnetically create food, computer designed buildings, clothing, and much more. Fossil fuels will hopefully become phased out sometime during the next few earth decades.

__Conclusions__

We should begin getting prepared for major changes in how we deal with all of earth's life forms, aside from harmful diseases, cancers, bacteria, and viruses that hurt humans. One of the ways that we could efficiently begin to make the world a better place once the replicator is in place, beginning a more powerful dollar, would to effectively change all major oil corporations business quickly as a result of making sure that gas stations obtain replicators then never buying gas from the fuel corporations ever again; or as often since they still may want fresh samples to replicate. The gas station owners will be able to sell gasoline for much less per gallon and still make a great percentage profit until the internal combustion engine is mostly phased out to be replaced by EM planetary vehicles and teletransportation over the following decades. All of humanity would start to profit on a personal level with a much healthier, cleaner planet Earth!

This can be a major change for all humanity to be able to communicate more freely in person, nearly as easily and efficiently as over the internet or phone. The ability to look another person in the eye while communicating can not be replaced by hearing their voice or seeing them on a computer screen. The eyes are the gateway to the soul. I truly hope that our long established governments can and will openly adjust to the larger universe as we now reside on a smaller planet in the greater scheme of things. All life needs may be able to be met without conflict over resources as a peace can be established through free trade of needs being met to help all to live more enjoyable lives, as we take pleasure seeing our families grow and prosper.

I truly hope and pray that our planet earth's governments and corporations will embrace the technologies of replicators, disintegrators, teletransporters, and EM vehicles to help all of humanity to begin to live at a lower cost and higher standard of living. I hope and pray that there will not be a corporate backlash to try to suppress such technologies, as they have the potential to evolve multibillion dollar industries like that of oil, medical treatment for profit, retail industries, and major corporate/government business interactions. I hope and pray that all of the governments and corporations of the world won't try to suppress the technological breakthroughs that will inevitably enable people to live more pleasurable lives without problems of hunger or major disease. Both the corporations and governments won't loose any power and their relative profits should increase. I also hope and pray that all people can be more accepting of each other, as we embrace each others differences to help give each other greater power in the areas in which we are ignorant. The atom, planetary systems, solar systems, galactic systems, the universe, and all levels of the universe are composed of positive and negative electromagnetic energy particles, waves, fields, and energy flow.

Ignorance is Bliss, But Knowledge is Power!